Dr. Earl Mindell's

What You Should Know About Fiber and Digestion

Other Keats titles by Dr. Earl Mindell

Garlic: The Miracle Nutrient
Live Longer and Feel Better With Vitamins and Minerals

See end of book for a complete list of titles in the "What You Should Know" series

Dr. Earl Mindell's

What You Should Know About Fiber and Digestion

Earl L. Mindell, R.Ph., Ph.D.

with Virginia L. Hopkins

Keats Publishing, Inc. New Canaan, Connecticut

Dr. Earl Mindell's What You Should Know About Fiber and Digestion is intended solely for informational and educational purposes, and not as medical advice. Please consult a medical or health professional if you have questions about your health.

DR. EARL MINDELL'S WHAT YOU SHOULD KNOW ABOUT FIBER AND DIGESTION

Library of Congress Cataloging-in-Publication Data

Mindell, Earl.
Dr. Earl Mindell's what you should know about fiber and digestion / by Earl Mindell, with Virginia Hopkins.
p. cm.
Includes bibliographical references and index.
ISBN 0-87983-745-4
1. Digestive organs—Diseases—Popular works. 2. Fiber in human nutrition—Popular works. 3. Digestion—Popular works. I. Hopkins, Virginia. II. Title.
RC806.M56 1997
613.2'63—dc21 96-46567
CIP

Printed in the United States of America

Keats Publishing, Inc.
27 Pine Street (Box 876)
New Canaan, Connecticut 06840-0876

98 97 96 6 5 4 3 2 1

Contents

INTRODUCTION

A Healthy Digestive System Is the Foundation of Good Health

In the first part of this book, I'm going to introduce you to the three basic parts of your digestive system in what may be a brand-new paradigm for you. I'm going to tell you that heartburn is usually caused by too little stomach acid; that your arthritis may be caused by microscopic holes in your intestines; and that your gas and bloating may have been caused by the antibiotics you took last spring. We're going to explore the stomach, its hydrochloric acid, and how to prevent heartburn; the small intestine, its digestive enzymes and food allergies; and the large intestine, its beneficial bacteria, and Candida yeast infections.

My approach to preventing and treating digestive difficulties may be brand new to you because, like so much of American life, the information you get from your doctor, from magazines and from TV about your digestive system tends to be dictated by large companies trying to sell you drugs rather than by the facts. Information based on advertising and marketing has very little to do with what's really going on in there!

This is why even though H2 blockers such as Tagamet and Pepcid are some of the best-selling drugs in the U.S., digestive problems remain one of the most persistent and bothersome of all American illnesses. If these drugs were working, wouldn't they be making you better? Instead, they cause a reliance on the drug to suppress symptoms. Taking H2 blockers does absolutely nothing to treat the *source* of indigestion; they only treat symptoms, and in the long run tend to make the problem worse and create dependencies and troublesome side effects. Your dependency on their drug just makes profit margins higher, and the side effects will be treated with yet another drug.

Poor digestion has repercussions in every other part of the body because it causes poor absorption of nutrients and puts stress on other organs in the body, as well as the brain, the blood, the lymphatic system, the muscles, ligaments and bones. It's amazing how many minor complaints, aches and pains will clear up when the digestion is cleared up.

Your key to good digestion is like the key to much else in life: moderation, balance, variety, and avoiding what is packaged and plastic. Overwork, overeating, underexercising, overmedicating and overconsumption of junk foods (which in my book includes nearly all processed foods) are some typical examples of imbalances that can lead to indigestion. Millions of Americans are plagued by indigestion, and yet good digestion and absorption of nutrients is one of the great keys to vibrant good health, clear-headedness and longevity. Take it from me—if your digestion isn't good, nothing else will be, either. You can eat the most nutritious diet possible, but if your body isn't absorbing the nutrients, it won't do you a bit of good.

I believe that delayed food allergies (also called food sensitivities or food intolerances) that damage the small intestines, and an overgrowth of Candida bacteria in the colon, rank among the leading causes of (or contributors to) such common chronic American illnesses as indigestion, hay fever, arthritis, headaches, fatigue, skin problems and autoimmune diseases. But most American doctors don't even acknowledge the existence of delayed food allergies or Candida overgrowth, let alone treat them. Why should they? They can't be quickly treated with a drug prescription, and they don't fit a neat pattern of cause and effect.

I have devoted the second part of this book to fiber. Why so much attention to a nonnutrient food substance? Because a lack of it can contribute to constipation, diarrhea, Crohn's disease, irritable bowel syndrome, colon cancer, varicose veins, excess estrogen and more! I'm going to fill you in on how fiber does its job, and how you can put more fiber into your life.

Looked at simplistically, your digestive system is one long tube, starting at your mouth and ending at your anus. The picture becomes more complex because each section of the digestive system is separated from its neighbor by a sphincter, or one-way valve. Your esophagus is separated from your stomach by the esophageal sphincter, which is opened by the action of swallowing a mouthful of food. Your stomach is separated from your small intestine by the pyloric sphincter, which allows just enough food through at a time to fill the duodenum, a sort of holding area between the stomach and small intestine. The esophageal sphincter and pyloric sphincter both protect delicate tissues from the stomach's powerful acid. In turn, the pyloric sphincter protects the stomach from the alkaline contents of the small intestine. Moving on down the line, the ileocecal sphincter separates the small intestines from the colon, protecting the small intestines from the bacteria of the colon. At the end of the colon, the small sphincter separates the colon from the outside world.

We're going to look at each of the digestive areas separated by sphincters—the stomach, the small intestines and the colon—because each has a distinct job to perform in moving food on its journey through the body.

Stick with me as we continue on this journey through your wondrous digestive system, implement my strategies for balancing it, system by system, and I can almost guarantee you'll create more energy, better memory, clearer thinking, and better overall health. You'll look better, feel better and do better. You'll also enjoy life more if you're not suffering from stomach cramps, bloating, gas and heartburn. Enough said!

PART I
Healthy Digestion Naturally

CHAPTER 1

The Stomach

Digestion doesn't really begin in the stomach, it begins when you smell food, or anticipate it. Those signals from your nose and your imagination stimulate the brain to send signals to the stomach to start producing digestive juices. Those in turn stimulate the pancreas and gallbladder, and thus the complex process of extracting nutrition from plants and animals begins.

DIGESTION BEGINS IN THE BRAIN

Your mouth "waters" in preparation for eating, and when that saliva mixes with the food you're chewing, salivary enzymes begin the process of breaking down the food. The most important of these is enzymes amylase, which breaks down starches.

While you're chewing, you're absorbing some fat-soluble nutrients through the mucous membranes in your mouth. These send further signals ahead to the digestive system about what types of digestive juices and enzymes to churn out. The better you chew your food, the more you're assisting the process of breaking it down into absorbable bits and pieces. (However, moderation being a key to good health, I don't recommend that you chew each mouthful of food 100 times as was advocated by a nutritionist in the 1950s—that's overdoing it!)

The esophagus is where peristalsis begins. This is the movement of special muscles that resembles kneading or milking—a constriction and relaxation that propels material through the digestive system.

THE STOMACH

The stomach is the biggest bulge in the tube that is the digestive tract, as most of us are well aware. But it is located higher than you might think, lying mainly behind the lower ribs, not under the navel, and it does not occupy the belly. It is a flexible bag enclosed by restless muscles, constantly changing form.

Virtually nothing is absorbed through the stomach walls except alcohol. An ordinary meal leaves the stomach in 3 to 5 hours. Watery substances, such as soup, leave the stomach quite rapidly. Fats take much longer to move through. Special glands and cells lining the stomach walls produce mucus, enzymes and hydrochloric acid, and a substance that enables vitamin B12 to be dissolved through intestinal walls into the circulation. A normal stomach is quite acidic, and gastric juice contains many substances.

Every day your stomach secretes hydrochloric acid (HCl) by the quart, along with an enzyme called pepsin which breaks down proteins. These secretions are stimulated by a hormone called gastrin which is released in the stomach in response to food. HCl is a potent acid, with the power to turn your streamed broccoli and grilled chicken into a wicked semi-liquid brew called chyme. HCl doesn't just break down foods, it also kills bacteria and parasites, allows the body to absorb minerals, and sets the stage for the absorption of B12 and folic acid in the small intestine. If the stomach doesn't contain enough hydrochloric acid, the pyloric valve into the small intestines won't open properly, causing food to sit in the stomach for hours and setting the stage for heartburn.

STOMACH ACID IS A FRIEND, NOT A FOE

If stomach acid is a friend of good digestion, why are we spending billions of dollars every year on antacids and H2 blockers such as Tagamet, Pepcid and Zantac, which block hydrochloric acid? Why do advertisements for these products blame all indigestion on stomach acid? It's because if your foods sits undigested in your stomach for hours, marinating in insufficient stomach acid, it can cause pain not only in the stomach, but from "reflux," or burping up of the contents of the stomach, which burns the lining of the esophagus—otherwise known as heartburn. Antacids and H2 blockers will temporarily block those symptoms, but in the long run will only make digestion more problematic and block the absorption of minerals and some B vitamins.

ARE YOU LOW IN HYDROCHLORIC ACID?

Most people with chronic indigestion or heartburn, and especially those over the age of 50, have low levels of HCl. One way to stimulate your digestive juices is to drink a glass of lukewarm water half an hour before eating. (Cold water suppresses digestive juices.) Other people swear by a tablespoon of apple cider vinegar in a glass of water before a meal. Vinegar is highly acidic and may provide your stomach with enough acidity for quicker, easier digestion. Instead of the vinegar you can try some type of bitters, such as the herb gentian or Angostura bitters, with warm water. A walk after a meal will help stimulate your digestive juices.

The most common symptoms of a stomach acid deficiency show up after eating in the form of heartburn, belching, bloating or a heavy feeling. If you can feel that most of your meal is still in your stomach more than 45 minutes after eating a normal meal, your stomach is working inefficiently.

If you suspect your stomach acid isn't up to par, and the other solutions I've given you aren't working, try taking betaine hydrochloride supplements. (Please don't take vinegar or start HCl supplements while you have an active case of heartburn. This will only irritate things even more.) Try taking one table of around 300 milligrams with food. You can increase your dose to two or more tablets per meal, but if you get a burning feeling in your stomach you're taking too much. (Some betaine hydrochloride supplements add pepsin, which is fine.)

A HOT REMEDY FOR SLUGGISH DIGESTION

Capsicum peppers—chili peppers and cayenne peppers—are some of the most versatile and powerful healing condiments known. The ingredient in peppers that makes them hot, capsaicin, is widely used in alternative medicine and in Asian medicine.

The hot-to-the-taste quality of peppers will initially produce a burning feeling in the mouth and sometimes on the skin, but the body then responds by blocking nerve pain transmissions, producing a pain-relieving effect.

Cayenne aids digestion by stimulating the production of stomach acid and increasing circulation in the gut.

There are many anecdotal reports of taking cayenne internally to stop internal bleeding and to treat ulcers. If you use cayenne to treat an ulcer, be careful not to take too much, and don't use it when the ulcer is inflamed. My preference would be that to treat ulcers you use it under the supervision of a health-care professional.

You can use cayenne and chili peppers in your food, or you can take them in capsules, which you can find at your health food store.

MY FAVORITE REMEDY FOR NAUSEA

Ginger (*Zingiber officinale*) is one of my favorite healing spices. The root of this plant has been used both as a food and as a medicine for thousands of years.

Ginger is best known for its effectiveness in relieving indigestion and most types of nausea, including morning sickness and motion sickness. In studies ginger has been shown to be more effective than dimenhydrinate (Dramamine) in relieving motion sickness. Researchers believe it is ginger's mild suppressing effect on the central nervous system that prevents nausea.

A substance in ginger called zingibain, which acts as a digestive enzyme, contributes to its ability to aid digestion and relieve gas. Ginger also stimulates the gallbladder to release bile, helping with the digestion of fats. And if that isn't enough, ginger may also help protect against ulcers by stimulating the production of protective mucus in the stomach. Ginger contains antioxidants, anti-inflammatory and antibacterial substances.

As well as using fresh ginger root, you can also take it in a powdered form in capsules, which you can find at your health food store.

THE REAL CAUSE OF ULCERS

For years mainstream medicine claimed that stress and excess stomach acid caused ulcers. Although stress can contribute to ulcers by either suppressing stomach acid or causing an oversecretion of it, we now know that the real culprit is a nasty spiral-shaped bacteria called *Helicobacter pylori* which lives in the stomach lining. This destructive little stowaway suppresses the production of hydrochloric acid, causes inflammation, reduces the protective mucus coating in the stomach, and creates holes in the lining of the stomach which allows the stomach acid to burn it and cause ulcers. It may be that in many people Helicobacter doesn't create enough damage to cause ulcers, but does suppress stomach acid enough to cause chronic indigestion and heartburn.

The standard medical treatment for Helicobacter infection is one or two antibiotics, usually tetracycline and amoxicillin, combined with a bismuth agent such as Pepto-Bismol, for a week to ten days. However, if you want to avoid antibiotics, you can try a combination of garlic capsules to kill the bacteria (one or two capsules with meals); licorice root tincture or capsules to aid in healing and protecting the stomach lining, and cabbage juice to speed healing.

Ulcers are also commonly caused by NSAIDS (nonsteroidal anti-inflammatory drugs) such as aspirin and ibuprofen. I encourage you to avoid them, especially if you're having stomach pains.

THE NATURAL WAY TO PREVENT AND TREAT HEARTBURN

If Helicobacter or low stomach acid isn't the cause of your heartburn, it most likely has a very simple cause. Most often the cause of heartburn is down-to-earth, simple and practical, and doesn't require any supplements or drugs.

Here are some of the most common causes of heartburn:

Eating in a hurry
Eating too much
Eating a lot of fried and fatty foods
Eating a lot of spicy foods
Eating foods that contain nitrates and nitrites
Eating citrus fruits (oranges, lemons, grapefruit)
Drinking a lot of coffee
Eating a lot of chocolate
Drinking ice-cold liquids before or with meals
Not chewing food thoroughly
Wearing tight clothing which constricts the stomach
Lying down after a meal, which irritates the esophageal sphincter

Once you get chronic heartburn, it tends to stay with you for the rest of your life. That's the bad news. The good news is that it can be managed by changing your lifestyle! Here are some other homespun tips for preventing heartburn:

Since your stomach has to work much harder to digest fat, it's important to avoid fried foods and stick to healthy oils such as olive oil and canola oil.

Eating small meals and chewing your food thoroughly is the next best prevention. Overeating and eating on the run are two of the most common causes of heartburn.

If you're overweight, losing some poundage

could be your ticket to digestive happiness (not to mention better overall health).

If you drink a lot of alcohol, cutting down (no more than two drinks daily) will almost certainly help long-term, and abstaining while you have symptoms will make healing much faster.

Are you one of those people whose stomach clutches when you're stressed? Try meditation, yoga, chi gong, Tai Chi, a hot bath or any form of relaxing exercise. (Flopping down on the sofa in front of the TV with a drink does *not* count as relaxation. Lying down will make it worse, alcohol will aggravate it, and TV is not deeply relaxing.)

Stop smoking and your heartburn may disappear. Nicotine relaxes the sphincter muscle that separates the esophagus from the stomach, allowing stomach acid to reflux (burp) up.

HOME REMEDIES FOR HEARTBURN

If you are suffering from heartburn, aloe vera juice and papaya are both effective heartburn remedies. You can find both, separately and in digestive formulas, at your health food stores. My grandmother drank a little bit of potato juice for heartburn, and swore by it.

At the first sign of heartburn, drink an eight-ounce glass of room-temperature water to rinse the esophagus of its acid bath and dilute your stomach acid if more comes up. (Ice-cold water interferes with digestion.)

Aloe vera gel soothes the esophagus and the digestive tract. It tastes bitter, but works well.

Take a teaspoon before a meal and a teaspoon after a meal.

If you get heartburn at night when you're sleeping, or are waking up with it, try putting a two-inch block of wood under each leg at the head of your bed. This will raise your chest slightly higher than your feet. If that's not practical, try sleeping on a wedge-shaped foam rubber pillow. Midnight heartburn happens because when you lie down and sleep, the sphincter muscle between the esophagus and stomach relaxes. If you have undigested food in your stomach, it's more likely to escape back up into the esophagus.

Baking soda will neutralize your stomach acid, temporarily relieving symptoms. Since this is a form of sodium, it's not recommended for those on a low-sodium diet, or for regular long-term use.

Herbal teas that can help relieve heartburn include fenugreek, slippery elm, comfrey, licorice and meadowsweet. They are all soothing to mucous membranes, and will help your esophagus heal. Drink lukewarm, with no lemon please! (High doses of comfrey tea used over a long period of time may cause liver damage, but I don't expect you to drink gallons of the stuff every day and it's one of nature's most soothing and healing substances.)

Herbal bitters are widely used in Europe as a digestive aid. The bitters stimulate the digestive juices. Europeans also eat a lot of the bitters such as endive and dandelion greens in their salads. Gentian root is the best-known bitter herb. Chamomile is also a "bitter" tea, although it doesn't have a bitter taste.

ANTACIDS WILL ONLY MAKE IT WORSE IN THE LONG RUN

Although antacids such as Mylanta, Rolaids and Tums can temporarily suppress the symptoms of heartburn, in the long run they'll do you more harm than good. You may even become dependent upon them. These over-the-counter medications help neutralize the acid in your stomach for up to an hour. That's fine for the moment, but your stomach may respond an hour later by producing even more acid to make up for what was neutralized, causing you to reach for more antacids. They also contain aluminum, silicone, sugar and a long list of dyes and preservatives. Your stomach acid is also one of your front-line defenses against harmful bacteria. Suppress it and the rest of your systems have to work overtime to protect you.

DRUGS THAT CAN CAUSE HEARTBURN

Many prescription and over-the-counter drugs can cause or aggravate heartburn. Drugs that specifically relax the esophageal sphincter muscle, allowing stomach acid to reflux up are anticholinergics (such as drugs to treat Parkinson's), calcium channel blockers (anti-angina drugs), nicotine and beta blockers (lower blood pressure and prevent spasms in the heart muscle).

Drugs That Can Cause Heartburn

Antibiotics
Cholesterol-lowering drugs
Blood pressure-lowering drugs
Asthma drugs
Corticosteroids
Antidepressants
Painkillers
Chemotherapy drugs
Tranquilizers
Barbiturates
Synthetic estrogens and progestins such as Provera and Premarin

CHAPTER 2

The Small Intestine

The small intestine is where most absorption of nutrients takes place. This is where the digestive system moves into high-performance mode, breaking down foods with digestive enzymes, extracting thousands of nutrients from the foods you eat, and sending them off to the liver for processing, which then sends them off to millions of jobs throughout the body.

The health of the small intestine is so important to your overall health, that naturopathic doctors estimate that some 60 percent of patients they see with previously unidentified symptoms (*i.e.*, those mainstream medicine could not help) are suffering, underneath all their other symptoms, from a dysfunctional gastrointestinal system.

A TOUR OF THE SMALL INTESTINE

Impossible as it may seem, you have about 22 feet of small intestine inside you with more than 2,000 square feet of surface area—that's about the square footage of an average two- or three-bedroom house! This relatively huge surface area is created by villi, tiny fingerlike protrusions in the intestines that interface between the intestines and the rest of the body, absorbing nutrients and sending them into the bloodstream where they are processed by the liver. When the intestines are damaged, the villi are damaged, paving the way for poor absorption of foods.

Unlike the stomach, which has an acidic environment, the small intestine has an alkaline environment, created by secretion of bicarbonate (like baking soda) from the pancreas. This alkalinity then stimulates the pancreas to secrete digestive enzymes. Meanwhile, the gallbladder releases bile, which aids in the breakdown of fats.

The liver, the gallbladder and the pancreas play an important role in the digestion of foods, so let's take a closer look at how they work.

The Liver

The liver is the main storage organ for fat-soluble vitamins such as A, D and E, and is also largely responsible for ridding the body of toxins. It is the largest solid organ of the body and weighs about four pounds. It is an incomparable chemical plant. It can modify almost any chemical structure for the body to use or eliminate. It is a powerful detoxifying organ, breaking down a variety of toxic molecules and rendering them harmless. It is also a blood reservoir and a storage organ for those fat-soluble vitamins, and for digested carbohydrate (glycogen), which is released to sustain blood sugar levels. It manufactures enzymes, cholesterol, proteins, vitamin A (from carotene), and blood coagulation factors.

One of the prime functions of the liver is to produce bile. Bile contains salts that promote efficient digestion of fats by detergent action, emulsifying fatty materials much as soap disperses grease when you're washing dishes.

The Gallbladder

This is a sac-like storage organ about three inches long. It holds bile, modifies it chemically, and concentrates it tenfold. The taste or sometimes even the smell or sight of food may be sufficient to empty it out. Constituents of gallbladder fluids sometimes crystallize and form gallstones. One of the best ways to keep your gallbladder healthy is to eat plenty of fiber.

The Pancreas

The pancreas provides important enzymes to the body. This gland is about six inches long and is nestled into the duodenum. It secretes insulin, which ushers sugar from the bloodstream into the cells. (Insulin is secreted into the blood, not the digestive tract). The larger part of the pancreas manufactures and secretes pancreatic juices, which contain some of the body's most important digestive enzymes, and bicarbonate, which neutralizes stomach acid.

DIGESTIVE ENZYMES

There isn't a cell in the body that functions without the help of enzymes. Enzymes are the magic ingredient that makes all of the other ingredients in the body work together. It is estimated that enzymes are facilitating 36 million biochemical reactions in the human body every minute. There are thousands, perhaps millions, of different enzymes at work, each with its own individual assignment.

Without the appropriate enzyme to bind to, vitamins are just so much organic matter, minerals are just so much inorganic matter, and even oxygen itself is just another molecule. Enzymes regulate all living matter, plants and animals alike. Take away enzymes and you no longer have something that is living.

Most enzymes are extremely tiny and found in very small quantities in the body. They work in organs, blood and tissue. The digestive enzymes, however, are a different story. Although you still need a microscope to see them, they are much larger than most other enzymes, and are present in the digestive system in large amounts.

Digestive enzymes are the catalysts in digestion and absorption, speeding up and enhancing the breakdown of foods. In one of those small miracles of biochemistry, the digestive enzymes cause biological reactions in our digestive systems without themselves being changed.

Food only becomes useful to the body after it has been converted to its parts: starches, sugars, amino acids, fats, vitamins, minerals and thousands of other nutrients such as phytochemicals from plants. Since each digestive enzyme works with a specific type of nutrient, a shortage or absence of even a single type of enzyme can make all the difference between health and sickness. One enzyme cannot substitute for another or do another's work.

We can also look at enzymes as the guide that shows the vitamin or mineral or fat the way into the cell. Without the introduction by the enzyme, the cell might never know the identity of the nutrient.

Some enzyme experts believe that factors such as stress, malnutrition, junk food, alcohol and cigarettes destroy and thus deplete enzymes. They theorize that many digestive problems and immune disorders happen when we are deficient in enzymes.

According to Ann Louise Gittleman, author of the book, *Guess Who Came to Dinner: Parasites and Your Health* (Avery Publishing, 1993), a lack of digestive enzymes also creates an ideal breeding ground for parasites. She explains that undigested food tends to rot and ferment in the intestines, which is the perfect environment for parasites.

GETTING TO KNOW YOUR ENZYMES

The enzymes that we know about are divided by what their purpose is in the body: they are called oxidoreductases, transferases, hydrolases, lyases, isomerases and ligases. The digestive enzymes are the hydrolases.

Digestive enzymes that end in *-ase* are named by the food substance they act upon. For example, the enzyme that acts on phosphorus is named phosphatase; one of the enzymes that works on sugar (sucrose) is called sucrase; and enzymes that break down proteins are called protease enzymes or proteolytic enzymes. Lipase breaks down fats, cellulase breaks down celluose, and amalase breaks down starches. Trypsin and chymotrypsin, produced by the pancreas, break down proteins.

The enzyme renin causes milk to coagulate, changing its protein, casein, into a form the body can use. Renin also releases minerals from milk, such as calcium, phosphorus, potassium and iron. Lactase is the enzyme that breaks down the milk sugar lactose. An absence of lactase is what causes many delayed allergies to milk.

HOW LIPASE WORKS

Enzymes that break down fat are especially important in Western cultures that tend to eat more fats than the body needs. Inadequate digestion of fat can cause stress in the entire digestive system, contributing to its chronic diseases.

Lipase and phospholipase break down fats in many stages, beginning with the upper portion of the stomach, called the cardial region. Here the lipase enzymes work in the acidic environment of the stomach to produce specific breakdown substances. If we aren't supplying enough enzymes here and in the main portion of the stomach to break down the fat we eat, when it reaches the small intestines it puts a much bigger load on the pancreas and gall bladder.

The lipases supplied by the pancreas only work in the alkalinity of the small intestines, producing a whole different set of fat breakdown products than the acidic environment of the stomach. An enzyme supplement can greatly aid the digestive system by making sure that fats we eat are well down the road to digestion by the time they reach the small intestine.

ENZYME PARTNERS

Although a digestive enzyme is a protein, it needs an amino acid and a cofactor, usually a vitamin or mineral, to work properly. Two of the most important digestive enzyme cofactors are magnesium and zinc. Magnesium alone is an essential cofactor (meaning it won't work without the magnesium present) for at least 300 different enzymes. Other mineral cofactors are iron, copper, manganese, selenium and molybdenum.

The B vitamins thiamin, riboflavin, pantothenic acid and biotin are all coenzymes that help us digest starches, fats and proteins. One study showed that taking a B complex vitamin supplement increased the activity of one enzyme by 25 percent!

Unlike the enzymes, the coenzymes are destroyed as they work with the enzymes. Thus, we need to replace our minerals and vitamins through what we eat. This is an important reason to take a good multivitamin every day (read my book in this series, *Creating Your Personal Vitamin Plan*).

KEEPING YOUR ENZYME TANK FULL

Many stresses of modern life can contribute to the destruction of enzymes, including toxins and pollutants; mental, emotional and physical stress; yo-yo dieting; drug and alcohol abuse; improper nutrition; and allergies. Some substances, such as fluoride, are necessary in extremely tiny amounts as enzyme cofactors, but in larger amounts they actually begin to destroy enzymes. Cadmium is found naturally with zinc, but when we get too much, it replaces zinc in the enzyme pathways and then can't finish the job, wreaking havoc on our cell membranes.

Some of the best food sources of enzymes are avocados, bananas, papayas, mangoes, pineapples, sprouts and the aspergillus plant.

Digestive enzyme supplements can work wonders for those who need a little extra help with digestion. If you have symptoms of indigestion such as gas, bloating, and cramping, or if you suspect you have food allergies, digestive enzyme supplements can help speed up the digestion process.

There are two sources of enzyme supplements: plants and animals. The most common sources of plant enzymes are papaya, from which papain is extracted, and pineapple, from which bromelain is extracted. Both papain and bromelain are proteases, or protein-digesting enzymes. I recommend you use plant-based enzymes.

When you take the digestive enzyme, be sure it includes the three major types of enzymes: amylase, protease (or proteolytic enzymes) and lipase. If you eat dairy products and want some help digesting the lactose in them, get an enzyme supplement that contains lactase. Take them just before or with meals.

THE FOUR BASIC TYPES OF DIGESTIVE ENZYMES

1. Amylase or amylolytic enzymes are found in the saliva, pancreas and intestines. They aid in the breakdown of carbohydrates.
2. Protease or proteolytic enzymes are found in the stomach, pancreas and intestines. They aid in the breakdown of proteins.
3. Lipase or lipolytic enzymes aid in the breakdown of fats.
4. Cellulase aids in breaking down cellulose.

AMYLASE ENZYMES THAT DIGEST STARCHES

1. Alpha-amylase is found in saliva and in the pancreas. It helps break down starches into sugars.
2. Beta-amylase is found in raw, unprocessed grains and vegetables, and also helps break down starch to sugar.
3. Mylase and glucomylase are starch-digesting enzymes capable of dissolving thousands of times their own weight in starches in the small intestine.

PROTEASE ENZYMES THAT DIGEST PROTEINS

1. Prolase is a concentrated protein-digesting enzyme derived from papain, which is extracted from papaya.
2. Protease is also extracted from papaya.
3. Bromelain is a digestive enzyme derived from pineapple.
4. Pepsin is released in the stomach, and splits protein into amino acids. In supplements, pepsin is made from animal enzymes.
5. Trypsin and chymotrypsin, produced by the pancreas, break down proteins.
6. Renin causes milk to coagulate, changing its protein, casein, into a form the body can use. Renin also releases minerals from milk, such as calcium, phosphorus, potassium and iron.
7. Pancreatin is an enzyme derived from the sections of an animal pancreas. This enzyme functions best in the small intestine.

ARE YOU ALLERGIC TO YOUR FOOD?

An estimated one in three people suffers from allergies of some kind. Allergies to pollen, dust, pets, manmade chemicals and food are the most common. One of the most common sources of digestive ailments is food allergies.

Before we go any further, let's qualify what I mean by food allergies. Allergy specialists love to argue about the definition of food allergies versus food intolerances or food sensitivities. For the purpose of simplicity, I divide all negative reactions to foods into two groups: immediate allergies and delayed allergies. This is easy because the treatment for delayed food allergies is the same, whether they are, strictly speaking, intolerances, sensitivities or allergies.

Immediate allergies cause symptoms such as hives, asthma, sneezing, watery and itchy eyes and nose, and even anaphylactic shock. Children suffer most from immediate allergies, and normally outgrow them, though a sudden stress combined with eating a childhood food allergen can bring it all back. Only one percent of the adult population is estimated to suffer from immediate allergies to foods. Seafood, strawberries, milk and beans are some of the more common foods that can produce immediate allergic reactions.

Delayed allergic reactions to food, also called food intolerances or food sensitivities, are far more common than immediate allergies. These symptoms may not show up for hours or days and can vary considerably. Many of our chronic complaints such as headaches, immediate allergies to pollens and chemicals, indigestion, stiffness and achy joints, and fatigue can all be symptoms of a delayed food allergy and the damage it does to the intestines.

When digestion becomes problematic due to chronic delayed food allergies, the rest of the body is more vulnerable to illness. For example, the inability to absorb nutrients can cause low-grade, generalized symptoms of malnutrition and essential fatty acid deficiency such as dull hair, dry skin and weak nails. Constipation can allow estrogen in the large intestine to be reabsorbed, causing PMS. The immune system becomes constantly stressed, causing greater susceptibility to colds and flu.

Many mainstream doctors do not even acknowledge the existence of delayed food allergies, but then again, they are unable to successfully treat many of the diseases caused or aggravated by them, such as Crohn's disease, psoriasis, irritable bowel syndrome (IBS), arthritis and autoimmune diseases. Your best bet if you have these illnesses is to go to a health-care professional who is able to recognize and treat them.

For most people, a delayed food allergy causes low-level chronic symptoms that worsen with age or under stress. Your energy may be low, you may have skin, hair and nail problems, minor aches and pains, headaches, various forms of indigestion, hay fever, stiffness in the joints, muscle weakness, and frequent colds. Many people think it's normal to walk around feeling like this, but if you fall into this category, removing one or two foods from your diet may change your life!

WHAT CAUSES DELAYED FOOD ALLERGIES?

The root causes of food allergies are interwoven with the root causes of intestinal damage. It's difficult to know whether intestinal damage causes allergies, or allergies cause intestinal damage. For example, if the stomach doesn't secrete enough hydrochloric acid, or the pancreas is unable to release adequate digestive enzymes, foods won't be broken down properly, which can damage the intestinal walls. This damage in turn reduces the absorption of nutrients and allows large, unidentified food particles to escape into the bloodstream, causing the immune system to react with inflammation (in effect, an allergy) somewhere in the body. In that case, the intestinal damage is causing the food allergy.

But suppose you have an inherited sensitivity to yeast, so that every time you eat bread, crackers, vinegar, beer and many other foods, and the yeast enters the bloodstream, your immune system responds with alarm. This may set up inflammatory reactions in the intestines, which cause intestinal damage. In that case the allergy has caused the intestinal damage. Fortunately, the treatment is the same, although if you have underlying genetic susceptibilities to allergies, you may want to focus more on strengthening your immune system and reducing inflammatory reactions. (For more on strengthening the immune system, see my book in this series, *Nutrition for Active Lifestyles.*)

I believe that much of what we called hyperactivity and treat with the drug Ritalin (one in 20 boys at least count—a national scandal) is really delayed allergic reactions to foods, especially food dyes and preservatives. These kids are set up for this by being on antibiotics for most of their toddler years, which severely suppresses the immune system as well as all the friendly bacteria in the large intestine.

By the time many kids are introduced to solid foods their immune system and digestive tract have been damaged by antibiotics, and their liver has been damaged by acetaminophen (children's Tylenol). This is a triple whammy! No wonder they're hypersensitive! Add to this almost constant exposure to environmental estrogens, known as xenoestrogens, through pesticide-sprayed fruits and vegetables, hormone-laced meat, and air pollution. Excessive estrogens cause sodium and water retention (which affects the brain, especially in children), inflammation, suppression of thyroid function, excess copper and many more unpleasant symptoms. A healthy digestive system might be able to excrete excess estrogen, but a damaged and constantly stressed digestive system will barely be delivering the necessary nutrients for growth and development.

Children who aren't breastfed, which gives them needed immunities, and who are introduced to hard-to-digest grains and other solid foods too early, tend to have more food allergies. Throughout our lives, but especially in childhood, the digestive system and immune system are busily sorting out what's friend and foe; what's a nutritious food, and what's a bacterial or viral enemy. Foods for which a child isn't prepared, or maybe even has a genetic susceptibility can be mistaken for an enemy, causing diarrhea, vomiting, asthma, fatigue, poor absorption or some type of inflammatory reaction such as eczema.

In adults, the underlying cause of food allergies can range from chronic stress which shuts down the secretion of digestive juices, alcoholism or poor diet, to many outside environmental factors that damage the intestines such as pesticides, parasites, heavy metals such as lead and mercury, food additives, and especially drugs such as antibiotics and NSAIDs (nonsteroidal anti-inflammatory drugs) that damage the intestines.

IDENTIFYING THE MOST COMMON FOOD ALLERGEN CULPRITS

The common delayed food allergens are wheat, corn, dairy products, soy, citrus fruit, the nightshade family of vegetables (potatoes, tomatoes, eggplant, red and green peppers, cayenne), peanuts (often caused by aflatoxins, a fungus found in most peanuts), eggs, beef and coffee—but we can become allergic to almost anything.

Irritable bowel syndrome (IBS) is often a delayed food allergy to dairy products. People with an allergy to gluten, known as celiac sprue, are allergic to all grains that contain gluten, including wheat, rye, barley and oats. Some of the worst and most insidious culprits, because they are hidden in processed foods, are food additives such as food colorings (especially red and yellow dye), BHT, BHA, MSG, benzoates, nitrates and sulfites—yet another great reason to avoid processed foods like the plague they are!

The most common symptoms of parasites, another cause of damaged intestines, are diarrhea and cramping. If you suspect you have parasites, which many people do after traveling in a foreign country, see a health-care professional for testing and treatment.

LEAKY GUT SYNDROME

Leaky gut syndrome occurs when the villi which line the intestines become damaged. The damage creates microscopic holes in the intestines, through which relatively large, partially digested food particles escape into the bloodstream. Bacteria and other microorganisms are also able to pass through the intestinal wall. The immune system detects these foreign invaders and sends out the troops to get rid of them. If this is happening constantly, such as every time you eat, the immune system will become overstressed and inflammation sets in.

Foreign particles in the bloodstream may migrate almost anywhere in the body, often to the place of greatest vulnerability such as the joints, where an inflammation reaction causes arthritic symptoms. A high percentage of arthritic symptoms can be effectively cleared up just by removing food allergens from the diet and healing the intestinal tract.

A leaky gut also interferes with the body's system for distinguishing safe and needed nutrients from foreign invaders. When digestion is proceeding normally, our foods are broken down into usable nutrients. As digestion moves along, special areas along the digestive tract identify the food or "tag" it as a friend, thus desensitizing it to the immune system. However, when the intestines are damaged, the tagging process starts to break down. When untagged food particles move into the bloodstream and run into the body's immune system, antibodies are made, to fight them and get rid of them.

So here we have the immune system on overtime trying to handle genuine foreign invaders such as bacteria and viruses; it's constantly going after large, partially digested food particles; and on top of it all, it's reacting to necessary nutrients as if they too were enemies!

Meanwhile, the liver, our most important detoxification organ which is already very busy shunting nutrients off to the rest of the body, is forced to work overtime to detoxify what the small intestines was unable to take care of.

PREVENTING LEAKY GUT SYNDROME AND FOOD ALLERGIES

Probably the single best way you can prevent leaky gut syndrome and food allergies is by avoiding anti-inflammatory drugs. My guess is that aspirin, ibuprofen and other similar anti-inflammatory drugs cause the vast majority of intestinal damage, which then causes delayed food allergies. We take these drugs casually, as if they are perfectly safe, but they really aren't. Aspirin alone is responsible for some 10,000 deaths and tens of thousands of hospitalizations every year due to intestinal bleeding. My advice to you as a pharmacist is to take drugs only when you really need them, and then for as short a time as possible, in as small a dose as possible. That advice alone could spare you a lifetime of illness!

Avoiding antibiotics whenever possible will also help prevent leaky gut syndrome and food allergies. As important as these miracle drugs are for treating life-threatening infections, they have serious side effects that have been largely ignored by mainstream medicine. Antibiotics suppress the immune system, and they kill both good and bad intestinal bacteria. As you'll discover in the chapter on the large colon, we need our friendly bacteria to stay healthy if we're going to stay healthy.

Fundamental to a healthy digestive system are my basic tenets for staying healthy: eating a wholesome, balanced diet free of processed foods; drinking six to eight glasses of clean water every day; getting some exercise at least three or four times a week; and taking a multivitamin every day.

I also want you to avoid drugs because many of them are very hard on the liver, which only adds to the burden on the immune system. It's always best to use natural alternatives to prescription drugs whenever possible.

Leaky gut syndrome can be made much worse when we aren't producing enough digestive enzymes to break down food particles in the intestines. Starting with the enzymes in the upper stomach, a snowballing effect can be produced, where food not digested properly at one stage puts a bigger load on the enzymes at the next stage, and so forth. One research study showed that 90 percent of people with gastric disturbances were not producing enough of one specific digestive enzyme, as compared to only 20 percent of a group of healthy people.

One of the best ways to combat leaky gut

syndrome is to take digestive enzyme supplements when you eat. These will help the digestive system break down the food particles better. Taking digestive enzymes can significantly increase the digestion of food, and greatly reduce symptoms of food allergies.

TREATING LEAKY GUT SYNDROME AND FOOD ALLERGIES

To treat leaky gut syndrome, you need to identify food allergens and take supplements to heal the intestinal tract. It's also important to support your liver by avoiding large amounts of alcohol and prescription drugs that stress the liver. To support the liver you can take the herb milk thistle (silymarin), which you can find where you buy your other supplements.

Some health professionals will test for leaky gut syndrome with a test called the lactulose/mannitol absorption test. The test measures levels of lactulose, which is made of very large molecules that normally don't enter the body. If the test shows elevated levels of lactulose it shows that large molecules are being allowed through the intestines.

AN ELIMINATION DIET MAY BE YOUR TICKET TO BETTER HEALTH

Tests that identify delayed food allergens vary in expense and reliability. The least expensive and in some ways the most accurate way to identify problem foods is to go on an elimination diet. If your self-discipline and your ability to track your symptoms every day are very good, you can try this on your own. Otherwise I recommend you do it with the support of a health care professional. I also recommend the book *Optimal Wellness* by Ralph Golan, M.D. (Ballantine Books, 1995), which contains a very thorough and detailed chapter on food allergies and elimination diets.

The basic principle of the elimination diet is to follow a specially limited diet for two weeks, and then reintroduce foods you suspect of being allergens one day at a time, while closely observing how your body responds. Your body will tend to have a strong response to an offending food after a few days of being off of it. Conversely, if you eliminate the offending foods during your two weeks, many of your chronic symptoms will disappear and you'll notice increased energy.

But before you start an elimination diet it's important to spend ten days keeping a daily journal of *everything* you eat and drink. That means food and drink, snacks and meals, eating in and eating out. At the end of the ten days, make a list of all the foods you ate every day, and another list of all the foods that you ate more than five times in that ten-day period. Odd as it may sound, the foods you ate every day are most likely the culprits. It may seem as if the foods you like best are those that make you feel best, but we tend to become addicted to the foods we are allergic to. You should be able to reintroduce most of your favorite foods back into your diet very gradually, although I don't recommend that you eat them every day.

Ideally you will eliminate both the daily foods and the "more than five times" foods from your diet for at least two weeks. If you suspect other foods of being allergens, also eliminate those. Sometimes supplements contain fillers and binders that are allergens, so eliminate those too for two weeks if they are suspect. (The label should identify all fillers and binder, or say that there aren't any. If it doesn't don't buy that vitamin.)

During the two weeks you are off your suspect food allergens be sure to keep a journal of the foods you're eating. People who are allergic to foods find that they tend to become allergic to almost *anything* that they eat every day. This makes it important to rotate the foods you do eat so that if at all possible you're not eating any one food more than every other day.

When you eliminate suspect foods, you must not eat even a tiny amount of them, or you'll throw off your test. Wheat, corn, dairy products and eggs are hidden in many processed foods, so if you're not familiar with something on a label, skip it. Your best bet is to eat whole, fresh foods that don't need a label!

If food allergy and a leaky gut are at the root of your problem, you should see a significant change in your health within the two-week period. Symptoms such as diarrhea and skin rashes should clear up within a week or so. Children tend to recover especially quickly from the symptoms of food allergies when they are put on an elimination diet.

People who have spent years with chronic low-grade health problems caused by food allergies and leaky gut syndrome are likely to be suffering from other problems such as Candida overgrowth, arthritis, autoimmune diseases and hay fever. In this case it's important to work with a health-care professional who can help you heal your whole body and bring it back into balance.

After two weeks, begin reintroducing foods, one every 24 hours. Keep a meticulous journal of symptoms. If you have any symptoms such as a faster pulse, rapid or irregular heartbeat, sudden fatigue, a feeling of heaviness or sleepiness, stomach cramps, bloating, gas, diarrhea, constipation, headache, dizziness, chills, sweats, flushing, skin rash, runny or itchy eyes or nose, stiffness, achiness, muscle weakness or any other unusual physical symptom, you are most likely sensitive to that food. Eliminate that food completely from your diet for two months, and then try reintroducing it again. If you are still sensitive to it, eliminate it for six months and try again.

Some foods may always be allergens for you, but if you're taking good care of yourself and your digestive system, you should be able to safely reintroduce food allergens to your diet. If you start eating them every day, you'll probably become allergic to them again. Children may have to outgrow a food allergy, which can be hard for everyone if it's a common food like wheat or dairy products.

SUPPLEMENTS FOR HEALING YOUR INTESTINES

Glutamine

Glutamine is an amino acid that is found in very high concentrations in the small intestines, and plays an important role in maintaining gut mucosa. Taking glutamine before and after surgery can significantly speed up healing time. Manganese is a mineral that is essential to the synthesis of glutamine, and glutamine is essential to the synthesis of niacin, as is tryptophan.

Glutamine may be a food for cancer tumors, so please don't take this as a supplement if you have cancer. I recommend 500 mg of glutamine taken three times daily with meals, when you're working on healing your intestinal tract.

Essential Fatty Acids

Essential fatty acids, and particularly GLA (gamma-linolenic acid) from borage oil or evening primrose oil, help prevent inflammatory reactions throughout the body, and specifically in the gut. GLA is an omega-6 fatty acid normally made in the body from the essential fatty acid linoleic acid.

The best way to insure that your GLA levels stay high and in balance with the other fatty acids is to avoid those things that deplete GLA. The biggest offender is the trans fatty acids found in hydrogenated oils used in margarine and nearly all processed foods. These "fake" oils rob the body of GLA. The second biggest offender in depleting GLA is processed foods depleted of GLA. Whole grains, and particularly oatmeal as well as many nuts and seeds, contain small amounts of GLA.

Taking too much alpha-linolenic acid, the omega-3 fatty acid found in flaxseed oil, is another way to suppress GLA production. Flaxseed oil is something of a nutritional fad right now, but I don't recommend you take large amounts of it long-term, as it is a highly unstable unsaturated oil that can do just as much harm as good if it's rancid or taken in excess.

You can also take GLA in the form of a supplement of evening primrose oil or borage oil. Follow the directions on the container.

It's also important to get plenty of omega-3 fatty acids, which are mainly found in fish oils. Eicosapentaenoic acid (EPA), found in fish oil, will also inhibit inflammation. This is why I encourage you to eat cold water, deep sea fish such as salmon, cod, sardines and flounder at least twice a week.

You should be taking fifty to a hundred more EPA than GLA oils. You only need 1–2 mg of GLA and 50–100 mg of EPA.

If you'd like a clearly written, well-researched book on prostaglandins and essential fatty acids, try *Enter the Zone* by Barry Sears (Regan Books, 1995).

HERBS FOR HEALING THE GUT

Licorice

You probably think of licorice as a candy because its distinctive flavor and sweetness make it a great sweet treat, especially for diabetics. It is commonly used in throat lozenges, and the Chinese give it to teething babies to help relieve pain and to young children to promote the growth of muscle and bones. Licorice is 50 times sweeter than sugar, but it is also a powerful medicine with literally hundreds of studies backing up its effectiveness.

Glycyrrhiza glabra (the Latin name for licorice) contains a chemical called glycyrrhizin which stimulates the secretion of the adrenal cortex hormone, aldosterone, and has a powerful cortisone-like effect. This makes it a useful anti-inflammatory. However, glycyrrhizin can cause high blood pressure, potassium loss and edema (swelling or water retention) due to sodium retention when it is used in high doses for many months. To avoid this effect, some licorice preparations are deglycyrrhizinized and known as DGL. People who are taking licorice for ulcers often take DGL, which comes in lozenge form because it needs to be chewed to be most effective.

The antiulcer effects of licorice are due to its ability to increase beneficial prostaglandins (hormonelike substances that regulate many bodily functions, including inflammation) that promote mucous secretion and the healing process in the stomach. According to a study cited by herbalist Donald Brown, licorice root, which stimulates growth of the surface layers of intestinal mucous lining, works even better when combined with ginger, which stimulates deeper layers of mucosal healing. In Europe licorice is often combined with the herb chamomile for treating stomach problems.

You can take licorice root in lozenges, capsules or tincture form, for up to two weeks, following the directions on the container. DGL lozenges are fine to take long-term. They should contain 250–500 mg of licorice and one to three of them should be chewed 15 minutes to half an hour before meals.

California Black Walnut and Yellow Dock

According to herbalist Michael Moore, the dried leaves of the California black walnut tree *(Juglans californica, J. hindsii)*, which is found throughout the U.S., can be made into a tea that is a wonderful healing agent to the small intestines. It acts as an astringent, a tonic, improves mucous membrane absorption, and reduces inflammation. According to Moore, you can use one part leaves to two parts water for the black walnut. Drink two cups of the tea daily.

For an even more effective remedy, add yellow dock *(Rumex crispus)* root, which you can find in tincture or capsule form at your health food store. Follow dosage instructions on the container.

CHAPTER 3

The Large Intestine

Your large intestine, also called the colon, is the waste disposal system of the digestive tract. The small intestine ends in a section called the ileum. This is where the body absorbs vitamin B12 and some of the fat-soluble vitamins, including A and E.

If you are facing another person, the colon is located in the navel area and is shaped something like an upside down U, with the small intestine ending at the bottom of the left side of the U. The ileocecal valve, a one-way sphincter, separates the small intestine from the large intestine. After food has moved through the small intestine, giving up its nutrients to the bloodstream, what's left is water, fiber, waste material such as bacteria, excess nutrients and undigested food. The ileocecal valve allows the waste material into a sac in the bottom of the colon called the cecum. Here the colon begins the process of pulling water and electrolytes (minerals) out of the waste material. The process continues as it moves up the ascending colon on the left side, and across the top part, called the transverse colon. As it moves down the descending colon it is formed into a more solid mass that is excreted through the rectum and anus as feces.

In a healthy body, it takes twelve to fourteen hours for waste material to make the circuit of the large intestine. Any material leaving the ileum and entering the cecum (where the small and large intestines join) should be quite watery. (If it isn't you will be constipated!)

KEEPING YOUR GOOD BACTERIA HEALTHY WILL KEEP YOU HEALTHY

The colon, in contrast to the germ-free stomach, is lavishly populated with bacteria, which are normal intestinal flora. These bacteria, also called probiotics, are also found in the mouth, the urinary tract and the vagina. There are about 100 trillion of these bacteria living in our bodies, and over 400 species. In a healthy body, the "good" bacteria run the show, and the bad bacteria are kept to a minimum.

When the bad bacteria outweigh the good bacteria, the result is often an overgrowth of bad bacteria (actually a fungal yeast) called *Candida albicans.* Women are familiar with the yeast infections that can occur in the vagina, but they can also happen elsewhere in the body, including the intestines. An overgrowth of yeast in the intestines can cause fatigue, bloating, gas, diarrhea, constipation and a long list of secondary symptoms such as headaches, mental fogginess, and pollen allergies.

Probiotics are the "good" bacteria. Your overall health is closely tied to the health of these bacteria. If they are sick, often so are you. They play a major role, along with our digestive enzymes, in digesting food and moving it out of the body.

The three most common families of friendly bacteria are called *Lactobacillus acidophilus, Lactobacillus bulgaricus,* and *Bifidobacterium bifidum.* These versatile bugs change and adapt rapidly, depending upon geographic location, individual biochemistry, and what types of unfriendly bacteria are invading the body at the moment. Probiotics are the ultimate antibiotics, elegantly crafted by nature to fight off unfriendly bacteria without killing the friendly ones. It's simple—take care of your friendly bacteria and they will take care of you.

Probiotics play other roles as well: your immune system depends on them; they manufacture the B vitamins and vitamin K; they reduce cholesterol, and help keep hormones in balance. A deficiency of probiotics can cause allergies, arthritis, skin problems, Candida, and may sabotage the role of your body's defense system in keeping cancer at bay.

The most common cause of a Candida overgrowth is taking antibiotics, which kill the friendly bacteria right along with the unfriendly ones, but don't seem to bother Candida. Always follow antibiotic treatment with at least two weeks of probiotics.

Steroids such as Prednisone and cortisone can also upset the balance of intestinal flora. Other factors are a poor diet, stress and poor digestion in the stomach and small intestine. Your friendly bacteria do better on the same diet that you do better on. Hmmm, what a coincidence. They like complex carbohydrates such as whole grains and beans, and fresh vegetables, and they don't like a lot of sugar, refined flour and dairy products.

Probiotics also decline as we age, so it's important to add probiotic supplements to your diet or eat unsweetened yogurt with live cultures (this is listed on the label) daily.

Many supermarkets and health food stores also sell "acidophilus," a milk product containing live cultures. Probiotic supplements are "alive" and have a relatively short shelf life of a few months. If you want to try probiotic supplements, please stick to the refrigerated capsules and reputable brands.

ARE YOU GETTING YOUR VITAMIN B12?

While we're on the subject of absorbing nutrients, I want to briefly touch on a very important topic—the absorption of vitamin B12—which happens at the end of the small intestine, in the ileum. A deficiency of vitamin B12 can actually cause the symptoms of senility and aging, such as an unsteady gait, memory loss, weakness, shortness of breath, indigestion, poor appetite, and for many, personality changes that include irritability, anxiety, depression and listlessness.

Most mainstream medical doctors will tell you that we get our vitamin B12 mainly from meat and dairy products, it's stored in the liver for future use, we need only a tiny amount to last us for years and years, and that therefore it's almost impossible to have a B12 deficiency, and B12 shots are a waste of money. They'll tell you that the only disease related to a B12 deficiency is pernicious anemia caused by a deficiency of intrinsic factor in the stomach. Unfortunately, by the time pernicious anemia shows up, a lot of damage has been done.

While it is true that we can store up to six years' worth of B12 in the liver, and that most Americans get plenty of B12 in their diet, it's also true that as we age our ability to absorb B12 can become so impaired that we do become deficient.

Without good digestion it's difficult to absorb B12, which is a complex process beginning in the stomach. There, parietal cells secrete hydrochloric acid, which releases the vitamin B12 from food, and they secrete a substance called intrinsic factor, which, with the aid of the pancreatic enzyme trypsin, binds to B12 in the ileum to carry it through the intestinal wall and into the rest of the body.

In other words, if your parietal cells are blocked, and your pancreas isn't secreting adequate digestive enzymes, your ability to absorb B12 can be greatly impaired. The process of aging tends to slow the action of parietal cells, but even bigger enemies of good digestion and absorption are a poor diet, antacid medicines, and the H2 blockers such as Tagamet, Pepcid and Zantac. Please throw away these medications! They may temporarily improve symptoms, but in the long run they will only hurt you.

The symptoms of B12 deficiency I mentioned above can begin showing up long before a blood test will show a deficiency. However, since the neurological damage done by a true B12 deficiency seems to be largely irreversible, it's not worth waiting around for it to show up that way. In a Dutch study of 16 older people with dementia, nine had normal blood serum levels of B12, but 12 had abnormally low levels of B12 in their cerebrospinal fluid (in the brain and spinal column). All of these patients improved significantly after receiving B12 injections.

If you or a loved one is showing symptoms of a B12 deficiency, I highly recommend you find a health-care professional who will give you 4–8 weeks of B12 injections to see if the symptoms improve. In the meantime, start using betaine hydrochloride and digestive enzymes, and at the end of the injections, start taking vitamin B12. It is not well absorbed when taken orally, so it's best to take it as a nasal gel or sublingually (under the tongue).

PREVENTING COLON CANCER

Colon cancer is a topic most people prefer to avoid, but 57,000 people will die of it this year. This is tragic, because so many colon cancers can be prevented so easily. For years I have been telling you that 68 percent of colon cancer is preventable.

We know you can have a genetic predisposition to get colon cancer, and we also know some major factors that can cause colon cancer:

1) There are a lot of toxic substances passing

through your system that will harm your bowel, and possibly create cancerous cells, if they sit there for very long.

2) Many studies have linked a high-fat diet to colon cancer. Fat stimulates your gallbladder to produce bile, which is one of those toxic substances that shouldn't sit there. Also, the high temperatures needed to cook fat can produce potentially cancer-causing substances in your food.

3) Low levels of the mineral selenium have been repeatedly linked to colon cancer.

FOUR WAYS TO PREVENT COLON CANCER

1) *Keep things moving through your bowels.* Chow down on vegetables such as broccoli, kale, carrots, onions, cabbage, collards, peas, potatoes and all dark green, yellow and orange vegetables. Why? They're high in fiber, so they keep your bowels moving, but that's not all. Vegetables contain substances that take toxins through your intestines quickly and harmlessly. And don't forget the wheat bran for more fiber. There is some evidence that fiber may even reverse the growth of precancerous polyps (bumps in the lining of your intestines).

2) *Cut down on the saturated fat you're eating.* Saturated fat is usually solid at room temperature. Examples are butter, the marbled fat in meat, the fat from fried foods, and hydrogenated oils. Your fat calories should be no more than 25 percent of your diet, and of that, 10 percent or less should come from saturated fat.

3) *Get your calcium*—it is paramount in the

prevention of colon cancer. The French eat five to six servings of yogurt daily, and even though they consume as much fat as most North Americans, the rate of colon cancer is much lower.

4) *Say yes to selenium!* Over the past few years a number of studies have linked low selenium levels with colon cancer. In a study at the University of Arizona, it was found that people with high levels of selenium in their blood had fewer colon polyps, which are often precancerous.

Selenium is an important part of an antioxidant enzyme called glutathione peroxidase. This enzyme may prevent damage to cells. Some studies show that it plays a role in the repair of DNA and helps activate the immune system.

Onion and garlic are a rich source of selenium. Other great sources of selenium are brown rice, seafood, kidney, liver, wheat germ, bran, tuna fish, tomatoes and broccoli.

You can take 100 to 200 micrograms of selenium daily.

RELIEVING CONSTIPATION NATURALLY

Constipation occurs when the bowels become hard and compact, and bowel movements are infrequent. If you aren't constipated yourself, you probably know someone who is. About 30 million Americans are afflicted with this uncomfortable and unhealthy problem.

So why aren't America's bowels moving? We need to eat more fiber, drink more water and get more exercise. Whatever you do, avoid those habit-forming over-the-counter laxatives.

When I was a kid and looking for something to do on a rainy day, my mother would sometimes give me some construction paper, some scissors, and then she would make me a paste out of white flour and water. It worked pretty well to stick things together! Well, that's just about what white flour does in your gut. It sticks to your ribs all right! If you don't believe me, get some white flour and add water until you get a nice Elmer's Glue-like paste. Your bowels need the fiber naturally found in whole grains, fruits and vegetables to function properly. The real secret to ending constipation is fiber, which you'll read a lot more about in Part II of this book.

If you don't drink enough water, your body won't have any to spare when it comes to making stools, and they will be hard and dry. This is painful and can cause bleeding and hemorrhoids. Exercise simply helps move the bowels—remember, moving your body will move your bowels!

It's also important to have a bowel movement when you have the urge to have one. Chronically holding back bowel movements can cause constipation.

Age can play a factor in constipation—the muscles we use to have a bowel movement can simply become weaker with time—especially if you've been stressing and straining them for a lifetime. Bran and prunes become a staple in the diet of many older people.

Some pharmaceutical drugs cause constipation, including pain killers, decongestants, narcotics, antihistamines, antidepressants and tranquilizers. Iron tablets can cause constipation. Dairy products, cheese in particular, can also cause constipation in some people.

Many of us become constipated when we travel—long hours of sitting, changes in diet and water, and a change in our daily routine can make us "irregular." Pack prunes or bran.

If the above remedies don't work to relieve your constipation, try psyllium powder. It's made from a fibrous plant that can help give your stools more bulk. It's the main ingredient in products such as Metamucil and Correctol, but you can get it in a much purer and cheaper form at your local health food store. Mix about a teaspoon with eight ounces of water or juice and drink it right away. It's very important to drink plenty of fluids when you're using a bulk-forming laxative such as psyllium.

Herbs such as cascara sagrada and senna can relieve constipation by stimulating the bowels. In general I don't recommend them because they can become habit-forming, but once in awhile they are fine. Cascara sagrada is made from the bark of a tree, and is fairly gentle. It is found in some over-the-counter laxatives, but I suggest you get it in capsule form at your local health food store. Senna is another plant laxative. It is a more powerfully bowel-stimulating laxative that has been commonly used for centuries. Stimulating laxatives become habit-forming if your bowels lose their natural ability to be stimulated. Try to use them only when absolutely necessary.

Magnesium can cause diarrhea, and thus can also be a remedy for constipation. This is particularly true when pregnancy is causing constipation. Taking 300 mg of magnesium is a safe and nutritional way to combat constipation. However, if you have chronic constipation, it's important to get to the cause and treat that.

My grandmother used to drink a glass of warm water before she went to bed at night. This makes sense, because many people are constipated simply because they don't drink enough water. Their stool becomes hard and compacted. Water naturally softens stools. If you are taking a diuretic medicine, it will tend to pull water out of your bowels, and can cause constipation. Please drink six to eight glasses of clean water every day.

Coffee can help move your bowels unless you overdo it. Drink too much coffee and you may end up constipated. I'm not a big fan of coffee, and I certainly don't want you to make it a daily habit, but if you find yourself constipated and without a handy remedy, you can almost always find a cup of coffee.

NATURAL REMEDIES FOR DIARRHEA

Diarrhea is loose, unformed, watery stools that tend to come frequently. Diarrhea is very common, especially in children. It can be caused by bacteria, such as in bad food; by a cold or flu virus; by parasites; by stress; by fatigue; by antibiotics and other prescription drugs.

Chronic diarrhea or diarrhea with blood or pus can be a symptom of a more serious illness and should be checked out by a health-care professional.

In the U.S. we talk about getting "Montezuma's Revenge," or diarrhea caused by drinking the water in a foreign country. However, when people from other countries come to the U.S., they also tend to get diarrhea because they aren't used to our bacteria, either! Any time you travel, it's important to stick to bottled water and eat plenty of yogurt to keep the colon healthy.

Here are some natural remedies for diarrhea:

Fiber

This may sound strange, but sometimes the cure for diarrhea is the same as the cure for constipation: fiber. Adding more bulk to your stools can give watery stools more firmness. Some people simply tend to have watery stools, and fiber is a good solution in that case.

Garlic

Studies have been done comparing garlic and drugs in their effectiveness for killing diarrhea-causing bugs. The garlic either wins or comes out performing as well as the drugs. It's cheaper than drugs, doesn't have any side effects and will also clean up parasites as it moves through. If you're using garlic to treat diarrhea you suspect is caused by a bug, it's best to take it in capsules so you don't further irritate your digestive tract. You can find many types of garlic supplements at your health food store.

Eat Bland Foods

It's important to eat simple, bland foods when you have diarrhea. In fact this is the one time when I would recommend white rice or white flour, because it is binding! Apples and bland cheeses also tend to be binding. Plain yogurt will help repopulate the friendly bacteria in the colon.

PART II
Fiber and Your Health

CHAPTER 4

Fiber and How It Works for You

For 20 years I have been telling people that if they would include more fiber in their diet a lot of their health problems would clear up. Cultures with plenty of fiber in their diet have virtually no constipation, no colon cancer, no varicose veins, no hemorrhoids, and on and on. Cultures that emphasize fiber-free refined grains and processed foods have all those illnesses and more. Americans seem to have to work to include 30 grams (g) of fiber in their daily diets. Members of cultures whose diet contains a lot of fiber may eat as much as 150 g in a day. That's a big difference! The average American consumes 10–15 g of fiber daily. I don't want you to try to eat 150 g of fiber a day—that's overdoing it—but I do want you to try to get 30–35 g daily.

My grandmother used to tell me I needed "roughage," but then it somehow fell out of favor and was just seen as a nutritionally unnecessary filler. But lately, fiber has become nutritionally chic. Mainstream medicine is catching on and catching up, and fiber has become the great white hope (pun intended), recommended by doctors and dietitians alike. It has finally dawned on our collective consciousness that our white flour, white rice and all the other white, refined foods in our diets are killing us, not just because they are "nutrition-free," but because they gum up the system. Fiber plays an absolutely essential role in good health, and that's what this part of the book is about.

WHAT IS FIBER?

Fiber is what gives plants their structure. It is the part of the food we eat that is not digested, passing through all the various chemical baths and enzymes of the gastrointestinal system untouched. It has virtually no calories, but provides bulk for digesting food, and especially for stools. Meat, eggs and dairy products do not contain any fiber.

SOLUBLE AND INSOLUBLE FIBER

Although fiber comes packaged in many different types of plants and their fruits, seeds, and nuts, the actual fiber itself can be divided into two types, soluble and insoluble. Soluble fiber becomes gummy in water, and insoluble fiber isn't changed in water. It is the insoluble fiber that passes through the digestive system unchanged, while soluble fiber has turned into a gum or jelly by the time it reaches the large intestine. Nature packages itself well when it comes to human health benefits, and fiber is no exception, since most fruits and vegetables come with both types of fiber.

Both soluble and insoluble fiber have many

benefits. Soluble types of fiber, including gums, pectins and mucilages, tend to stabilize blood sugar, reduce cholesterol and blood pressure, and provide a friendly environment for "good" colon bacteria. Insoluble types of fiber, which include cellulose, hemicellulose and lignin, add bulk to the stools, prevent constipation, help remove toxins from the bowel, and keep the intestines clean by their scrubbing action.

SOLUBLE FIBER

Sources: vegetables (especially onions), fruits, oat bran, gums from nuts, seeds, beans

Benefits: Stabilizes blood sugar, lowers cholesterol and blood pressure, provides a friendly environment for "good" bacteria.

INSOLUBLE FIBER

Sources: whole grains (wheat, barley, rye for example), vegetables, beans.

Benefits: Adds bulk to the stools, prevents constipation, helps remove toxins from the bowel, and keeps the intestines clean by having a scrubbing action.

ALL FIBER

Reduces fat intake by increasing volume of food and giving a full feeling. Speeds up transit time.

SUPPLEMENTAL SOURCES OF FIBER

- Wheat bran
- Oat bran
- Rice bran
- Prunes and figs
- Flax seeds
- Apple pectin
- Legumes
- Psyullim
- Guar gum and other gums
- Whole grains

INCREASING TRANSIT TIME

You already know that fiber increases the bulk of stool, which helps relieve constipation, but it can also increase transit time, which is the amount of time it take for food to go through the entire digestive process. Transit time will vary a lot just from person to person, based on genetics, environment and personality, but the amount of fiber in the diet also plays a significant role. Cultures that eat a lot of fiber tend to have much faster transit times than cultures with a low fiber intake.

However, if you are one of those people with too fast a transit time (that tends to also come with loose, watery, unformed stools), then fiber may also normalize your transit time by adding bulk to the stools and absorbing water.

Soluble fiber eaten without insoluble fiber may speed up transit time, which for most Americans is a positive benefit.

Ideally in most people, transit time should vary from 12 to 18 hours. For most Americans, transit time is 50–60 hours. You can find out what your transit time is by eating some corn or swallowing three or four 10-grain charcoal tablets (found at your local pharmacy) and tracking how long these substances take to show up at the other end. Corn tends to show up in feces unchanged, and charcoal will turn the stools black.

Oral contraceptives tend to slow transit time, and excess alcohol consumption speeds transit time.

FIBER AND DISEASE

Constipation

Clearly one of the biggest benefits of eating a high-fiber diet or taking fiber supplements is relief from constipation. Fiber is by far the best remedy for this common affliction. However, if you increase your fiber up to 30–35 g daily and you still have constipation, see a health-care professional to make sure there isn't something more serious going on.

Diverticulitis/Diverticulosis

Diverticula are pouches in the wall of the large intestine. Constipation and straining can force fecal matter into these pockets, causing them to enlarge, to become inflamed, and to leak digestive matter into the surrounding tissues. Since whatever is in the large intestine is full of noxious bacteria and waste matter, this causes infections and abscesses in the tissues surrounding the leaky diverticula.

Digestive matter that gets caught in the diverticula sits there and ferments, rots and becomes toxic to your system. Environmental toxins and waste products can also accumulate.

In diverticulosis, the diverticula bulge outward from the intestine, but there isn't inflammation. This is a common ailment of older people and can cause gas, pain in the navel area, especially on the left side, and sometimes bleeding. Diverticulosis, if left untreated, can cause diverticulitis.

The best remedy for diverticulosis is to avoid processed foods and make fiber-rich foods a staple of the diet.

The best type of supplemental fiber to take for diverticulosis is psyllium, since it has a fairly gentle action on the bowel but adds plenty of bulk.

Diseases of Excess Estrogen

The ailments caused by excessive estrogen, coined "estrogen dominance" by John R. Lee, M.D., author of *What Your Doctor May Not Tell You About Menopause* (Warner Books, 1996), are many. I'm going to mention some of the most common here, but I recommend that if you're a woman over the age of 40, you read his book—it's a real eye-opener.

Excess estrogen can contribute to insulin resistance, block thyroid function, and lower zinc levels, which impairs immune system function. This is a good hormone to have in balance!

Lack of fiber causes excess estrogen to be recycled back into the body through the large intestine. A diet with plenty of fiber will soak up excess estrogen and other waste material in the large intestine, and it will be excreted. When there isn't enough fiber to act like a sponge for the waste matter, or when there is constipation and the waste matter has to sit in the bowel for longer than it should, estrogen gets recycled back into the system rather than being excreted.

PMS (Premenstrual Syndrome)

PMS tends to have multiple causes, but balancing excess estrogen can bring quite a lot of relief to many women. The hormone progesterone balances, or opposes, estrogen. If a woman is deficient in progesterone at the end of her cycle, which may happen if she doesn't ovulate, she will have too much estrogen relative to not enough progesterone, and will have estrogen dominance symptoms such as water retention, irritability, sore breasts, headaches and fatigue. Adding supplemental fiber to the diet at the end of the menstrual cycle can help greatly to reduce estrogen levels and thus PMS symptoms.

There have been many studies showing that fiber lowers estrogen levels. In one, twelve premenopausal women ate a daily diet consisting of 30 percent of calories from fat, and 15–25 g of fiber, for one month. Then, for two months they ate a daily diet consisting of only 10 percent fat and 20–35 g of fiber. At the end of the very low-fat diet, estrogen levels had fallen significantly. (By the way, I do not recommend you reduce your fat to 10 percent of calories—that's too low, and it's not healthy. A more gradual approach of 20–25 percent calories from fat and 30–35 g of fiber daily will achieve the same result over a longer period of time.)

Breast Cancer

Breast cancer is an estrogen-driven cancer, and women at risk for it should strive to keep their estrogen levels in balance. Getting plenty of fiber is an essential part of a breast cancer prevention plan.

A study of women with benign breast tumors compared to healthy women found that those without breast tumors ate significantly more fiber.

The other benefit of eating plenty of fiber-rich vegetables is that many of them have phytohormones that take up estrogen receptor sites on cells, thus in effect blocking the action of the estrogen and lowering its levels in the body.

Menopause Symptoms

Menopause is not an estrogen deficiency disease, and adding estrogen to the body of a menopausal woman, especially one who is overweight, is asking for problems. Again, I recommend Dr. Lee's book mentioned above, and that you keep your fiber high when you are menopausal, to keep your hormones in balance.

High Cholesterol

Although high cholesterol is vastly overrated as a risk factor for dying from heart disease, it is a good idea to keep it under control. You don't want sky-high cholesterol because that's a sign that something is not working right in your cardiovascular system.

We know from many scientific studies that fiber plays a direct and substantial role in lowering cholesterol. Soluble fiber in particular seems to keep cholesterol levels in the healthy range.

The Johns Hopkins Medical Institutions did a study of fiber and cholesterol in China, and showed that eating oatmeal or buckwheat not only lowers cholesterol, it also lowers blood pressure. An amount as small as a one-ounce serving eaten daily can give the blood pressure-lowering effect.

Another study of 60 people with moderately high cholesterol who ate just 20 g of fiber a day for nine months found that their ratio of LDL ("bad") to HDL ("good") cholesterol improved by 11 percent, and their LDL cholesterol dropped nine percent. Best of all, these changes started showing up after only three weeks of adding fiber to the diet.

High Blood Pressure

As I mentioned above in the section on high cholesterol, a Johns Hopkins study in China showed that eating buckwheat and oatmeal—as little as an ounce a day—significantly lowered blood pressure. That sure beats beta blockers!

Heart Disease

We know that fiber lowers high cholesterol and high blood pressure, both risk factors for heart disease, but a high-fiber diet seems to protect against heart disease even without factoring in blood pressure and cholesterol levels.

One major study kept track of more than 43,000 male health professionals between 40 and 75 years of age for six years. Those who had the most fiber in their diets had half as many heart attacks as those who had the least amount of fiber in their diets, regardless of their fat intake, exercise and smoking. That's some potent medicine! I don't know of any prescription drug that even comes close to that kind of track record for preventing heart disease.

Although all types of fiber were protective, those from breakfast cereals were the most beneficial.

Colorectal Cancer

Fiber so clearly prevents colon cancer that every box of bran cereal should be able to make that claim, in big bold letters.

A study published in the *Journal of the National Cancer Institute* put 411 patients who had benign tumors in their colon on a diet that reduced fat by 25 percent and added 25 g of wheat bran every day, as well as 20 mg of beta-carotene (which is an insignificant amount). Patients on this regimen had no more adenomas after two to four years. Zero! Now there's a cheap, easy way to avoid surgery of the rectum.

If you are at risk for colorectal cancer, in addition to fiber, you should be getting 200 mcg of selenium daily, taking plenty of antioxidants, and eating little to no red meat.

Irritable Bowel Syndrome (IBS)/Colitis

Colitis is a catchall word for many types of inflammation of the large intestine. One of the most common is irritable bowel syndrome (IBS). People with IBS tend to have sharp pains in the navel area and alternating diarrhea and constipation. Since people with colitis very often have a delayed food allergy, please read the chapter on the small intestines and how to detect food allergies. People with IBS tend to be allergic to wheat or gluten, so avoid wheat bran if you have IBS unless you're certain you're not sensitive to it.

Insoluble fiber is often irritating to people suffering from a colitis attack, but soluble fiber can be quite beneficial. Insoluble fiber can be very gradually introduced when symptoms aren't present, and that will help prevent a recurrence and keep the gastrointestinal tract cleaned up.

Diabetes

Anything that slows the entry of sugar into the blood can be helpful to those with diabetes. Soluble fiber can slow the emptying of the stomach and coat the small intestine, thus slowing the release of sugar into the bloodstream and reducing the need for large amounts of insulin at one time. This is why it's so much more beneficial to eat an apple than drink apple juice.

Varicose Veins

Varicose veins have become twisted and so large that they can no longer efficiently carry blood. While obesity can certainly aggravate varicose veins, one of the major causes is straining during a bowel movement. Good bowel habits at an early age that prevent constipation and straining can prevent unsightly varicose veins later in life! Fiber is good preventive medicine in more ways than one.

Hemorrhoids

Hemorrhoids are enlarged veins in the rectum and anus, usually caused by constipation and straining. Yet another reason to eat plenty of fiber!

Gallstones

The gallbladder produces a substance called bile, which is squirted into the small intestine to emulsify and digest fats. Gallbladder pain can radiate up to behind the shoulder blades and under the right collarbone. One of the surest remedies for this type of pain is to take a fiber such as psyllium for a few days, and avoid fatty foods.

Gallstones, which are hardened deposits of calcium salts and other components of bile, can be prevented by eating plenty of fiber. Now isn't it easier to eat your bran cereal every morning than to have surgery to remove a gallstone?

Obesity

Fiber can help with weight reduction, for the simple reason that it is nearly calorie-free yet provides bulk, giving a feeling of being full. Again, an apple is more filling than apple juice, and an orange is more filling than orange juice.

INCREASING YOUR FIBER INTAKE

The best way to increase your intake of fiber is to eat more whole grains, as well as fresh fruit and vegetables, and cut down on the amount of white flour and other processed foods you eat. This will keep your bowels moving, and keep toxins from accumulating. It's always preferable to get your fiber from your food first, and from fiber supplements such as psyllium or bran second.

It's important not to introduce too much fiber or "roughage" into your diet too quickly. If you do, you may suffer from bloating, gas and abdominal cramps. Gradually introduce more whole grains, more fresh fruits and vegetables, and more water to your diet. If you're not getting some exercise every day, add that too. Give it three weeks. If that doesn't work, try bran or prunes. Start with a heaping tablespoon of bran and then gradually add more if you need to.

Another reason to increase your fiber very gradually is that as you begin to move the accumulated junk out of your intestines, a lot of toxins can be released, sometimes causing tiredness and headaches. If you do it gradually, and drink plenty of clean water (at least six to eight glasses daily), you will minimize these effects. You can also take milk thistle (the active ingredient is silymarin) to support your liver as it helps the body clear out the toxins.

I think that the best all-around fiber food is wheat bran, but as always, moderation is the key. Overdo it and you'll lose the benefits and gain new digestive problems. You'll get bran naturally if you eat whole grains. There are many cereals in the supermarket with bran in them, and they usually advertise it in big letters! You can buy bran in a jar and sprinkle it on your cereal or in your yogurt—try starting with a heaping tablespoon. Wheat bran is the most common bran in America, but rice and corn bran also work to relieve constipation.

Fiber-rich prunes and figs work wonders for many people with constipation. Have three or four for breakfast and see what happens. If they are too chewy for you, soak them in water overnight.

Popcorn is a good laxative and fiber for some people.

Beans are a particularly beneficial food for many reasons, including a high fiber content. They range in fiber content from 4 to 7 g per half-cup serving. They are rich in soluble fiber that lowers blood cholesterol and blood sugar. Beans also contain insoluble fiber, which can help with constipation.

To get the fiber benefit from wheat, you

need to look for "whole grain wheat" on the label, rather than just "whole wheat."

When you're eating fruits and vegetables, include the skin when possible.

When you're looking for a fruit treat, always go for the whole fruit instead of the fruit juice which has little to no fiber. Mother Nature's packaging is optimal for good health, and that includes fiber content.

BENEFITS FROM FIBER

Better digestion
Better elimination
Lowers cholesterol
Lowers blood pressure
Prevents colon cancer
Prevents appendicitis
Prevents breast cancer
Reduces PMS symptoms
Reduces menopausal symptoms
Weight loss
Detoxification
Healthy gallbladder
Stable blood sugar

FIBER-FULL FOODS

Fruits
Vegetables
Brown Rice
Dried beans
Wheat bran
Barley
Rye
Oats
Seeds
Nuts
Popcorn

SOME OF MY FAVORITE SOURCES OF FIBER (in grams)

1 cup chestnuts 14.5
½ cup oatmeal 7.7
1 cup lentils 6.4
corn on the cob 5
½ cup brown rice 5.5
5 halves dried pear 4.97
1 large artichoke 4.5
2-½ figs 4.45
1 small potato 4.2
1 cup raspberries 3.69
½ cup black cherries 2.95
1 apple 2.8
½ avocado 2.8
1 cup kidney beans 2.78
1 papaya 2.3–2.35
1 cup lentil sprouts 2.3–2.35
1 cup carrot juice 2.3–2.35
1 pear 2.3–2.35
½ cup sunflower seed 2.25
½ a guava 2.2
½ cup peanuts 1.94
½ cup sesame seeds 1.8

REFERENCES

Alberts, D S., et al., Randomized, Double-Blinded, Placebo-Controlled Study of Effect of Wheat Bran Fiber and Calcium on Fecal Bile Acids in Patients With Resected Adenomatous Colon Polyps, *Journal of the National Cancer Institute,* January 17, 1996;887(2):81–91.

Bagga, D, et al., Effects of a Very Low Fat, High Fiber Diet on Serum Hormones and Menstrual Function, *Cancer,* December 15, 1995;76(12):2491–2496.

Baghurst, PA, et al., Dietary Fiber and Risk of Benign Proliferative Epithelial Disorders of the Breast, *International Journal of Cancer,* 1995;63:481–485.

Barnes, RMR, IgG and IgA Antibodies to Dietary Antigens in Food Allergy and Intolerance, *Clinical and Experimental Allergy* 1995:25(Suppl. 1):7–9.

Bengmark, S, Jeppsson B., Gastrointestinal surface protection and mucosa reconditioning, *J Parenter Enteral Nutr* 1995;19:410–415.

Blomqvist, BI, et al., Glutamine and Alpha-Ketoglutarate Prevent the Decrease in Muscle Free Glutamine Concentration and Influence Protein Synthesis After Total Hip Replacement, *Metabolism* 1995;44:1215–1222.

Buchman, AL, Glutamine: Is it a Conditionally Required Nutrient For Human Gastrointestinal System? *Journal of the American College of Nutrition,* 1996;15(3):199–205.

Burr, ML, et al., Effects of Changes in Fat, Fish, and Fibre Intakes on Death and Myocardial Reinfarction. *Lancet* 1989;2:757–761.

Cara L, et al., Effects of Oat Bran, Rice Bran, Wheat Fiber, and Wheat Germ on Postprandial Lipemia in Healthy Adults, *Am J Clin Nutr* 1992;55:81–88.

Carter, C, Dietary Treatment of Food Allergy and Intolerance, *Clinical and Experimental Allergy* 1995;25(Suppl 1.):34–42.

Eaton, K.K., et al., Gut Permeability Measured by Polyethylene Glycol Absorption in Abnormal Gut Fermentation as Compared With Food Intolerance, *Journal of the Royal Society of Medicine,* February 1995;88:63–66.

Fehily, AM, et al., Diet and Incident Ischaemic Heart Disease: The Caerphilly Study, *Br J Nutr* 1993;69:303–314.

Giovannucci, E, et al., Intake of Fat, Meat, and Fiber in Relation to Risk of Colon Cancer in Men, *Cancer Research,* May 1, 1994;54:2390–2397.

Hallfrisch J, et al., Diets Containing Soluble Oat Extracts Improve Glucose and Insulin Responses of Moderately Hypercholesterolemic Men and Women, *Am J Clin Nutr* 1995;61:379–384.

He, J, et al., Dietary Macronutrients and Blood Pressure in Southwestern China, *Journal of Hypertension,* 1995; 13(11):1267–1274.

Humble, CG, et al., Dietary Fiber and Coronary Heart Disease in Middle-Aged Hypercholesterolemic Men. *Am J Prev Med* 1993;9:197–202.

Hunninghake, DB, et al., Hypocholesterolemic effects of a dietary fiber supplement. *Am J Clin Nutr* 1994;59: 1050–1054.

Hunninghake, DB, et al., Long-Term Treatment of Hypercholesterolemia With Dietary Fiber, *The American Journal of Medicine,* December 1994;97:504–508.

Jenkins, DJ, et al., Effects of Blood Lipids of Very High Intakes of Fiber in Diets Low in Saturated Fat and Cholesterol, *N Engl J Med* 1993; 329:21–26.

Kaaks, R, et al., Dietary Fiber and Colon Cancer, *Path. Biol* 1994;42(10):1091–1092.

Khaw, KT, Barrett-Connor E., Dietary Fiber and Reduced Ischemic Heart Disease Mortality Rates in Men and Women: a 12-year Prospective Study, *Am J Epidemiol* 1987;126:1093–1102.

Kromhout, D, et al., Dietary Fiber and 10-year Mortality from Coronary Heart Disease, Cancer and All Causes: The Zutphen Study. *Lancet* 1982;2:518–521.

Lanza, E, et al., Dietary Fiber Intake in the U.S. Population, *Am J Clin Nutr* 46:790–797,1987.

Lessof, M.H., et al., Reactions to Food Additives, *Clinical and Experimental Allergy,* 1995;25(Suppl. 1):27–28.

Li, J, Langkamp-Henken, B, et al., Glutamine prevents parenteral nutrition-induced increases in intestinal permeability, *J Parenter Enteral Nutr* 1994;18:303–307.

MacLennan, R, et al., Randomized Trial of Intake of Fat, Fiber, and Beta-Carotene to Prevent Colorectal Adenomas, *Journal of the National Cancer Institute,* December 6, 1995;87(23):1760–1766.

Marlett, JA, Content and Composition of Dietary Fiber in 117 Frequently Consumed Foods, *J Am Diet Assoc* 92:175–186, 1992.

Probert, CSJ, et al., Some Determinates of Whole-Gut Transit Time: A Population Based Study, *Quarterly Journal of Medicine* 1995;88:311–315.

Rainsford, KD, Leukotrienes in the Pathogenesis of NSAID-Induced Gastric and Intestinal Mucosal Damage, *Agents Actions,* 1993;39 (Special Conference Issue) C24–C26.

Rimm E, et al., Vegetable, Fruit, and Cereal Fiber Intake and Risk of Coronary Heart Disease among Men, *JAMA* 1996;275:447–451.

Ripsin, CM, et al., Oat Products and Lipid Lowering: A Meta-Analysis, *JAMA* 1992;267:3317–2235.

Smith, U., Carbohydrates, Fat, and Insulin Action, *Am J Clin Nutr* 1994;59 (Suppl):686S–689S.

Stoll, BA, et al., Can Supplementary Dietary Fiber Suppress Breast Cancer Growth?, *British Journal of Cancer,* 1996;73:557–559.

Swain, JF, et al., Comparison of the Effects of Oat Bran

and Low-Fiber Wheat on Serum Lipoprotein Levels and Blood Pressure, *N Engl J Med* 1990;322:193–195.

Terho, EO, Savolainen, J, Diagnosis of Food Allergy, *European Journal of Clinical Nutrition,* 1996;50:1–5.

Trowell, HC, Burkitt DP, eds., *Western Diseases: Their Emergence and Prevention.* London: Edward Arnold Publishers, 1981.

Warner, JO, Food Intolerance and Asthma, *Clinical and Experimental Allergy* 1995;25(Suppl. 1):29–30.

Warner, JO, Food and Behavior, *Clinical and Experimental Allergy* 1995;25(Suppl. 1):23–26.

Watson, W, Food Allergy in Children, *Clinical Reviews in Allergy and Immunology* 1995;(13):347–359.

Willett, WC, et al., Dietary Fat and Fiber in Relation to Risk of Breast Cancer: An 8-Year Follow-up, *JAMA* 1992;268:2037–2044.

Wynder, E, et al., High Fiber Intake: Indicator of a Healthy Life-Style, *JAMA,* February 14, 1996;275 (6):486–487.

RECOMMENDED READING

Gittleman, Ann Louise, *Guess Who Came to Dinner: Parasites and Your Health,* Garden City, New York, Avery Publishing, 1993

Golan, Ralph, MD, *Optimal Wellness,* New York, Ballantine Books, 1995

Lee, John R. MD, *What Your Doctor May Not Tell You About Menopause,* New York, Warner, 1996.

Mindell, Earl, *Earl Mindell's Anti-Aging Bible,* New York: Simon & Schuster, 1996.

Mindell, Earl, *Earl Mindell's Food as Medicine,* New York: Simon & Schuster, 1994.

Mindell, Earl, *Parent's Nutrition Bible,* Carson, Cal: Hay House, 1992.

Mindell, Earl, *The Mindell Letter,* a monthly newsletter published by Phillips Publishing, Potomac, MD., 1-800-787-3003.

Sears, Barry, *Enter the Zone,* New York, Regan Books, 1995.

INDEX

Dr. Earl Mindell's

What You Should Know About . . .
series

Beautiful Hair, Skin and Nails

Better Nutrition for Athletes

Fiber and Digestion

Herbs for Your Health

Homeopathic Remedies

Natural Health for Men

Natural Health for Women

Nutrition for Active Lifestyles

The Super Antioxidant Miracle

Trace Minerals

22 *Ways to a Healthier Heart*